Quantum Physics 101: An Introduction To Quantum Physics And The Origins

Table of Contents

Introduction to Quantum Physics

Quantum Physics or otherwise known as the Quantum mechanics is a science that relates to very small bodies that are tiny in nature. This is a scientific principal that gives an explanation of matters/substances behavior as well as their interaction with the energy as atom as well as subatomic particles. This guide has been designed in a step by step order giving all the relevant information that is required to introduce you into Quantum Physics and their origins as a beginner to the Quantum mechanics.

Origin Of Quantum Physics

Before the 19th century classical physics was the only science which was used to explain energy and matter in small scale that could be familiar to the human experience. This was in inclusion of the astronomical bodies. Although this science still remains the key to measurement in the modern technology and science towards the end of 19th century there was a discovery by scientists in both the large and the small bodies, these are the macros and the micros. In them the phenomena of classical science could not work. This limitation of classical theory leads to revolution in physics where the theory of relativity and Quantum

Physics were introduced to fill the gap. The remaining steps in this guide explain Quantum Physics after its discovery in early 20th century.

Definition Of The Word "Quantum"

In this sense and the whole of this guide "Quantum" refers to the minimum quantity of any physical substance that is involved in any type of interaction. Taking the consideration of light, it behaves like waves in some respects and particles in some respects. This is due to the fact that even matter-particles for instance atoms and electrons have got some wave like behaviors. A neon light gives only a very small amount of light. Quantum Physics exhibits the fact that all the forms of electromagnetic radiations and light comes in very small units that are referred to as photons which predicts the colors, spectral and energy intensity of the light.

The First Quantum Theory: Max Planck And Black-Body Radiation

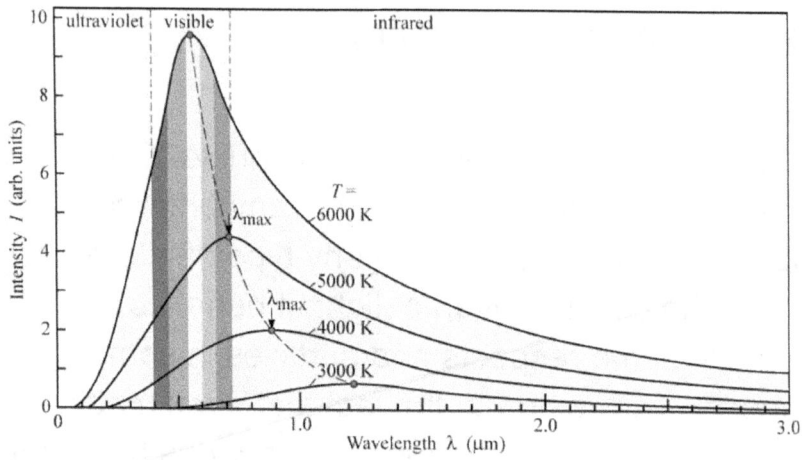

Thermal radiation is the form of electromagnetic radiation that a given object emits from its surface due to the temperature present in it. When we heat a given object in a way which is sufficient it will in turn emit light which is red at the end, this is due to the fact that the body becomes red hot. When it is heated further there is a series of color change that is from Red- to Yellow- to White- then Blue. This series of color change is due to the fact that shorter wavelength light spectra are emitted, they contain frequencies which are higher. This in turn gives the fact that in emitter which is perfect is also a good absorber. This is due to the fact that such an object will become totally black when cooled. In this case it will have absorbed all the light that falls on it but emits none. A "Black body" is an ideal thermal emitter. Such a body emits "Black-body radiation".

Max Planck came up with the first theory which gives explanation to the thermal radiations full spectra- in 1900. In this case he developed a mathematical model where thermal radiation and a given set of harmonic oscillators were in equilibrium. The energy of each oscillators were "quantized"- an assumption made to show that each oscillator produces a given amount of energy. According to him the Quantum of energy to each oscillator was related proportionally to the oscillator's frequency. The constant of that was arrived at the proportionality is the "Plank Constant". This is written as "h". This constant has the value 6.63×10^{-34} J s. Planck's Law was the initial and first Quantum theory to be introduced in physics.

Photons: The Quantization Of Light

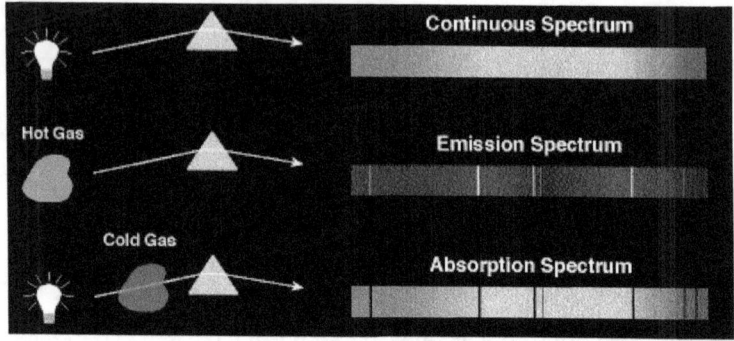

This was a forward step taken by Albert Einstein in the year 1905. With him, he gave the suggestion that quantization did not just involve mathematical trick. On it he added that it also involved the beam of light energy that is in the individual packets- currently referred to as photons. Thus, the energy of a single photon will be given by the product of the frequency of the energy and Planck's constant.

In the 19th century, light was considered to be flowing in a wave and this was the result of behaviors of light like polarization, diffraction as well as refraction. Actually according to James Clerk Maxwell magnetism, light and electricity are all manifested by the same phenomena; which is the electromagnetic field. He explains light as waves which constitute a combination of magnetic fields as well as oscillating electric. The Einstein's "Photon model" came into place when it was able to explain the photoelectric effect successfully. This effect has been explained in the next step.

The Photoelectric Effect

$$E_{photon} = h\nu$$

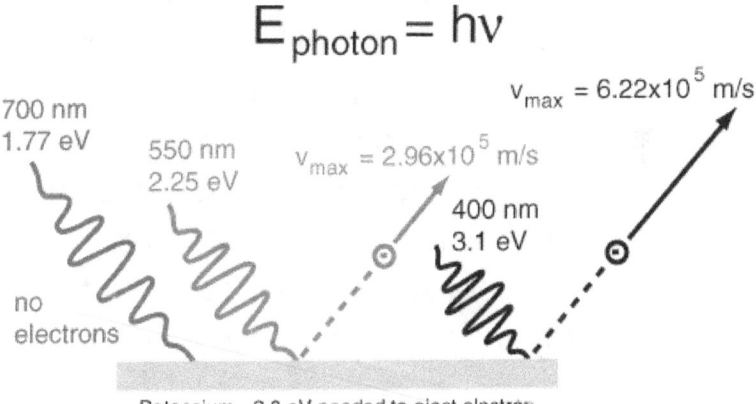

700 nm
1.77 eV

550 nm
2.25 eV

$V_{max} = 2.96\times10^5$ m/s

$V_{max} = 6.22\times10^5$ m/s

400 nm
3.1 eV

no electrons

Potassium - 2.0 eV needed to eject electron

Photoelectric effect

Back in the year 18887, Heinrich Hertz made an observation that light could successfully perform ejection of electrons from a metal. Again, Philipp Lenard, in the year 1902 made the discovery the highest possible energy in relation to an electron that has been ejected will depend with the light's frequency and never its intensity. This explains why there can be a case where electrons will never be ejected even with very high intensity. Threshold frequency is the light's lowest frequency required to emit an electron, it is different from metal to metal.

According to the Einstein's explanation, he argued that a beam of light has got the photons, which are particles stream and also a frequency "f". The energy present in that photon will be equal to "hf". This implies that there is no any effect on energy that relates to the beam's intensity. He further explained that in order to remove an electron from a given metal, "work function" which is a certain energy amount is required- it is denoted by "φ". With his further explanation, when the work function is higher than the photon's energy there will be no sufficient energy that is required to remove the electron from the given metal.

With his description that also argued that light is composed of particles gave an extension of the Planck's notion. This is the notion of energy that is quantized- whereby more or less amount of energy can be delivered by a given photon depending on its frequency. There was a compromise on the particle state of light due the fact that is was explained that light also hade waves. This resulted to the consequences of quantization of light.

Consequences Of Quantization Of Light.
The reason why infrared radiation and visible light can't cause sunburn whereas ultraviolet light can

cause sunburn is due to the energy and electromagnetic radiation frequency of individual photon in the light. An ultraviolet photon has got a very high energy which is enough to result into cellular damage. In the case of infrared its photon has got very low energy that can only warm the skin, without giving any sunburn. Einstein rejects the wave depended particle approach (classical) on the favor of particle based analysis. In the latter case each and every particle's energy is absolute and varies with its frequency in small steps- in this case the energy is quantized. This gives the implication that there is identical energy in all photons of similar frequency.

This gives the implication that the energy that is emitted by a given forged piece of iron or start per unit time depends on the photons emitted in the unit time as well as the energy amount carried by the same photons. In any case despite the fact that there is variation of the energy that is imparted by photons in any given frequency, it is important to understand the fact that in the photoelectric device the initial state of energy of the respective electrons before the absorption of light isn't uniform. Thus, we can only talk of quantization of light when referring to a mass of vast number of electrons and particles and not a single electron or photon. In so doing we will overcome the changes that might have appeared to the photon due to absorption of illuminations of low frequencies.

Matter Quantization: The Bohr Model Of The Atom

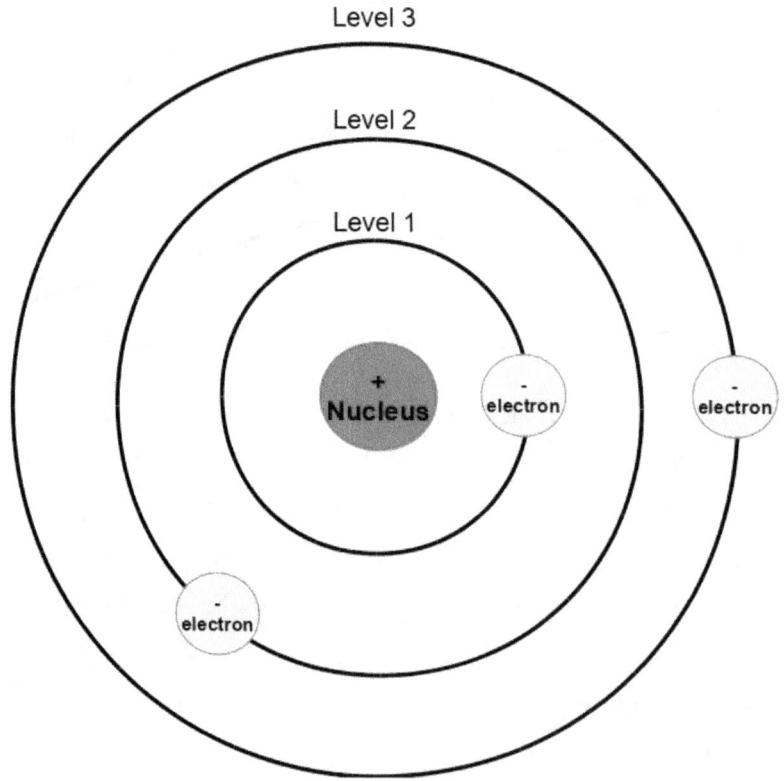

Early 20th century the classical theorists argued that there is a centripetal acceleration on all orbiting electrons. This implied that they also give off electromagnetic radiation. In the ear 1913 a scientist and physicist by the name Niels Bohr came with a proposal of a new model of the atom. In this model there was the inclusion of the orbits of quantized electrons. He argued that electrons still orbits the nucleus just like how the sun is orbited by the planets; although for the case of the electrons they will only be permitted to use certain orbits and not to orbit to any other distances. This implied that when a given atom absorbed or emitted energy that electron could not move from one orbit to another in a continuous trajectory. This is as opposed to how the classical theory argued.

According to Bohr the electrons jumped from one orbit to another. In this case it gave off the light emitted in a form of photon. The energies that are emitted by the photon were highly dependent on the differences in energy that was present between the orbits. In the first place there were very many critiques on the Bohr's model. Many argued that this model was wrong, although at the end it was evident that the model was good to suit the quantum physics. With his explanation Bohr argued that matter has also got some wave-like properties. According to him, an electron beam can also exhibit "diffraction". This is a similar case just like the beam of light or a wave of water. Thus small molecules and atoms have got the same phenomenon. To proof the above a double spit experiment was undertaken- this experiment is explained in the next step.

Double-Slit Experiment

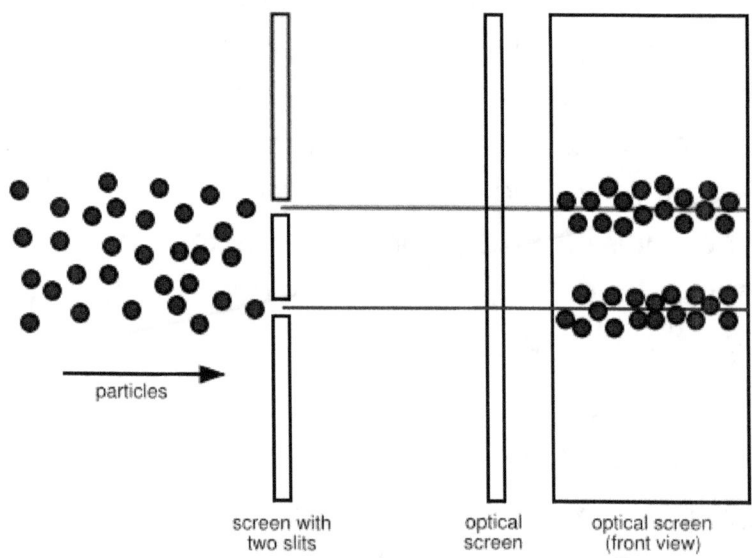

This experiment was performed in the by Augustine Fresnel and Thomas Young back in the year 1827. In it a light beam was directed through closely spaced slits that were very narrow (two of them). This produced light's interface pattern that had dark band on the screen. In case one of the slits was covered we would expect the fridges to be halved due to the interference. This was due to diffraction. It was termed as a simple diffraction.

This experiment was also performed with molecules, atoms, as well as electrons where there was the same type of interference. With it was evident that all matters have a possession of both wave and particle characteristics. This was also evident when only a single particle was allowed to pass through the slit at a time. Thus, it can e summed up that the quantum particle have got wave characteristics when passing through slits which are double. With this experiment the quantum complimentarity was arrived at, which states that, "a quantum particle will behave like a particle when a particle experiment is carried and like a wave when wave experiment is carried.

Application Of The Bohr Model

Another scientist by the name "De Broglie" came with an expansion of the Bohr's model. He argued that there were wave-like properties to all the electrons which orbited the nucleus. According to him he argued that for an electron to be observed there was the need of a situation which permitted the wave to stand around the nucleus. For instance, when a sting is tied in to ends tightly and made to vibrate, it will be very hard to know it is a string is a faster oscillation, it can only be known as a string when it's made to stand when all the particles present in it will remain still. This is known as the "Violin sting" experiment.

De Broglie came up with the suggestion that the only electrons which orbited the nucleus were the ones whose orbits circumference had an integer number of the wavelengths. This gave the implication that the wavelengths of the electrons gave the determination of the Bohr orbits from the nucleus. Bohr radius is the smallest possible distance from the particles nucleus.

Why Quantum Physics Was Developed

Due to the act that results from different experiments could not be explained by the classical physics there was a reason to come up with a more vivid and pleasant theory or way to give explanation to such experiments. Galileo Galilei as Isaac Newton was responsible for the classical physics theories. According to the classical argument were that electrons orbited around the nucleus in way that resembles how the sun is orbited by the planets. Actually in the normal world such a phenomena could result to crush in the atom or the particle due to the many and compact electrons in it.

Their argument about the dependency of chemistry in the atoms' electrons interaction and dependency of life on chemistry was all incorrect. Actually there were more than thousands of experiments that could not be explained by the classical physicists. This led to the then scientists coming up with new ideas to offer explanation to the atomic level of these phenomena. Perhaps, it is incorrect to say that the classical physics is wrong.

The only thing to put across is the fact that there are very many flaws in the classical physics more so to the atomic level of particles. This was what called for more experiments explanations that lead to the Quantum Physics. The latter also offers explanation to phenomena that relates to very fast electrons like the speed of light in which there is relativity taking place. For the things that are large than atoms as well as very slow when compared to the speed of light, classical theory offers great explanation that work out well. It is also very easy to use such law in than Quantum Physics on the given areas, due to the very many and extensive mathematics involved in the latter.

Importance Of Quantum Physics

The following are some of the areas in which Quantum Physics becomes an important thing. They are the areas where Quantum Physics can explain but the classical physics cannot:

- Energy discreteness
- Duality of Wave particle of matter and light
- Quantum tunneling
- The Heisenberg uncertainty principle
- Particle spin

All these have been explained in the next steps vividly, to offer a clear outlook of what they are comprised of.

Energy Discreteness

This can be evident by looking at the light spectrum which is emitted by atoms that are energetic. For example, taking an instance of the sodium vapor street lights where orange- yellow light is emitted you will note that there are different color of lines which are individual in the beam of light. The lines are representation of energy levels which are discrete. According to this explanation that has been given vividly in the initial instances electrons can only be in existence in discrete energy levels. This is what offers prevention in them from spiraling in the nucleus. It is also predicted by the classical physical but to lights of slower speed. In this case some of the properties which are atomic are quantized giving Quantum Physics its name.

Duality Of The Wave-Particle Of Matter And Light

This is a phenomenon that explains a situation where light behaves like particles in experiments done to study the particle nature of light, whereas in experiments designed to study the wave nature of light, it behaves like waves. This gives the implication that light both behave like wave and particles at the same time. In the classical experiments and theory there were no any explanations to support the fact behind this. This is due to the fact that it was not a good idea to say that light was made up of particles that moved up and down in the form of a wave. Although when the Quantum Physicists can they explained the fact that matter and particles are in existence in the form of particles. The reason as to why there is a wave-like behavior in light is due to the fact that there are closely placed particles in the beam of light that are very small and accumulated together.

Quantum Tunneling

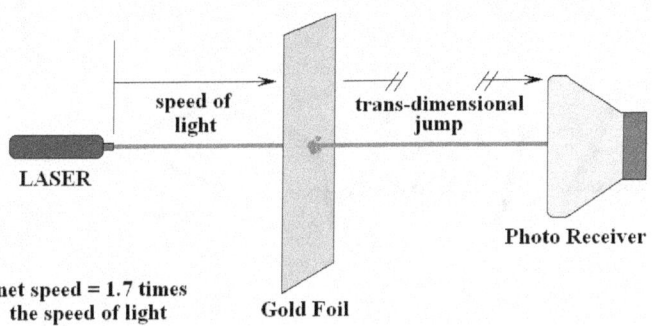

Photon Quantum Tunneling

In the Quantum Physics this is one of the most interesting phenomena. Actually this is the reason as to why computer chips were invented; essentially computers would have been as large as a room. Normally the probability of where a particle will be is determined by a wave. There is reflection of the wave when a barrier is encountered, although there might be some small portions which might lead to the barrier. This is the tunneling quantum. It is explained by the fact that in the occurrences of energy drop in particles there will be a back reflection of the particles.

The Heisenberg Uncertainty Principle

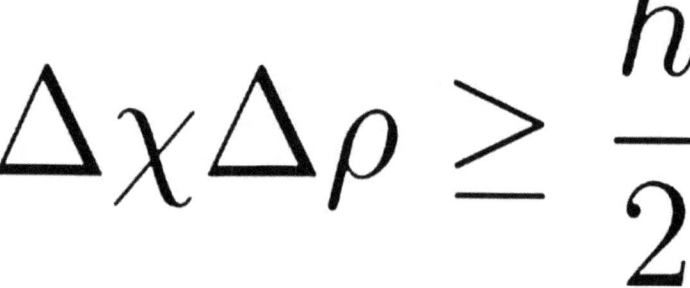

$$\Delta x \Delta \rho \geq \frac{\hbar}{2}$$

Generally at the atomic scale it is very delicate to perform measurements. This is due to the fact that one cannot just have a tape measure and take the measurement of a single electron successfully. This scientist by the name Heisenberg was the first one to come up with the realization that there were intrinsic uncertainties in certain pairs of measurements. His idea fills the flaw that was in the classical physics. He came up with the idea that measuring up to scale and accuracy of billionth of a centimeter was impossible, which was a great compliment to the Quantum Physics. Actually according to him it could be not possible to come up with a device that can measure such a scale

Spin Of A Particle

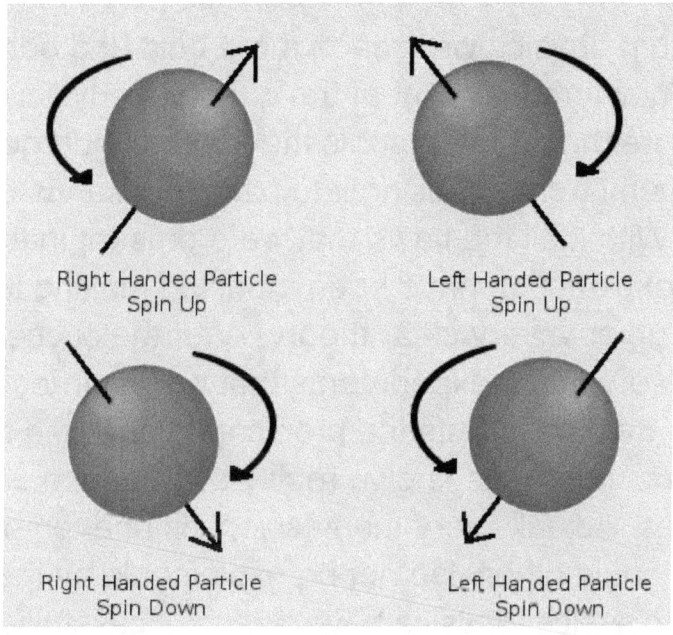

Quantum Physics has also been applied in the spin of particles. In the year 1922 a scientist by the name Otto Stern came up with an experiment hose results were not explainable by the classical physics. In it atomic particles were in possession of angular momentum which was intrinsic or spinning, whose spin could actually be quantized. This meant that there were discrete values in the particles. With it this spin has been explained vividly in the Quantum Physics whereas the classical physics cannot explain it.

Quantum physics was made more bold and bolder by every new predicament of the quantum physicists. For more than 80 years this theory has undergone very many fronts, making it to pass thousands of tests. With it all the flaws that were present in the initial explanation have been modified making it to come up as a very clear theory. After the successes in the very many experiments that were carried out by the then physicists the predicative power in this theory came to the known reality of Quantum Physics. Actually very many scientific phenomena which could otherwise not be explainable by the use of the classical physics have been successfully explained by the Quantum Physics. This has seen it a success.

The weakness of the Quantum Physics and mechanics only arises at the point where it offers explanation of the world on two ways and not in any single way. According to it an object is represented as to whether it can be observed or not observed. This make every physicist to come up with his or her own way when using the theories in the Quantum Physics in the two cases- that is when in object is not observed and when it is observed. Although due to the fact that this theory can be applied to all things including the smallest atoms particles and the biggest objects it still hold its importance in physics and stands a great chance to offer solution to the very many experiments that are done in the current technological world. This success in the field of Quantum Physics has played a great role that is important to some of the many applications of the ideologies in the physics world. Some of these applications have been explained in the next and last step in this guide.

Application Of Quantum Physics

Quantum Physics has been wide applied in very many areas in the modern world. This includes in the transistors, laser, the electron microscope as well as magnetic resonance imaging. There is also a special Quantum Physics mechanics that has its application in macroscopic quantum phenomena. These are in inclusion of the super-conductors as well as super-fluid helium.

It is in the study of such semiconductor that has led to inventions of transistors and diodes. The two have a great role in the modern electronic. Quantum tunneling is also a great science in making of small tools like the light switch. This is due to the fact that it was evident that the oxide layer could not penetrate light. Other things that have been made though the use of Quantum Physics science includes the memory chips that are evident in USB drives where quantum tunneling has been applied in erasing of the memory cells which are present in them.

www.ingramcontent.com/pod-product-compliance
Lightning Source LLC
Chambersburg PA
CBHW070308190526
45169CB00004B/1546